MAY 1 0 2006

P9-BVG-898

Jack the Builder

by Stuart J. Murphy • illustrated by Michael Rex

HarperCollins*Publishers*

LEVEL
1

Glenview Public Library
1930 Glenview Road
Glenview, Illinois

To our very own Jack the Builder!
—S.J.M.

To Fern and John, a teacher and a builder
—M.R.

The publisher and author would like to thank teachers Patricia Chase, Phyllis Goldman, and Patrick Hopfensperger for their help in making the math in MathStart just right for kids.

HarperCollins®, ☙®, MathStart® are registered trademarks of HarperCollins Publishers. For more information about the MathStart series, write to HarperCollins Children's Books, 1350 Avenue of the Americas, New York, NY 10019, or visit our website at www.mathstartbooks.com.

Bugs incorporated in the MathStart series design were painted by Jon Buller.

Jack the Builder
Text copyright © 2006 by Stuart J. Murphy
Illustrations copyright © 2006 by Michael Rex
Manufactured in China by South China Printing Company Ltd.
All rights reserved

Library of Congress Cataloging-in-Publication Data
Murphy, Stuart J.
  Jack the builder / by Stuart J. Murphy ; illustrated by Michael Rex. — 1st ed.
    p. cm. — (MathStart)
  "Level 1."
  ISBN-10: 0-06-055775-3 (pbk.) — ISBN-10: 0-06-055774-5
  ISBN-13: 978-0-06-055775-1 (pbk.) — ISBN-13: 978-0-06-055774-4
    1. Counting—Juvenile literature. 2. Blocks (Toys)—Juvenile literature. I. Rex, Michael, ill.
II. Title. III. Series.
QA113.M8844 2006   2005002663
513.2'11—dc22   CIP
                                                AC

Typography by Elynn Cohen 1 2 3 4 5 6 7 8 9 10 ❖ First Edition

Be sure to look for all of these **MathStart** books:

Blocks, *blocks*, BLOCKS!

Jack has a great big pile of blocks!

He can stack them, pile them, build them high.

Just two blocks can build . . .

. . . a robot toy to play with.

Jack has

He'll put one more on top.

2

3

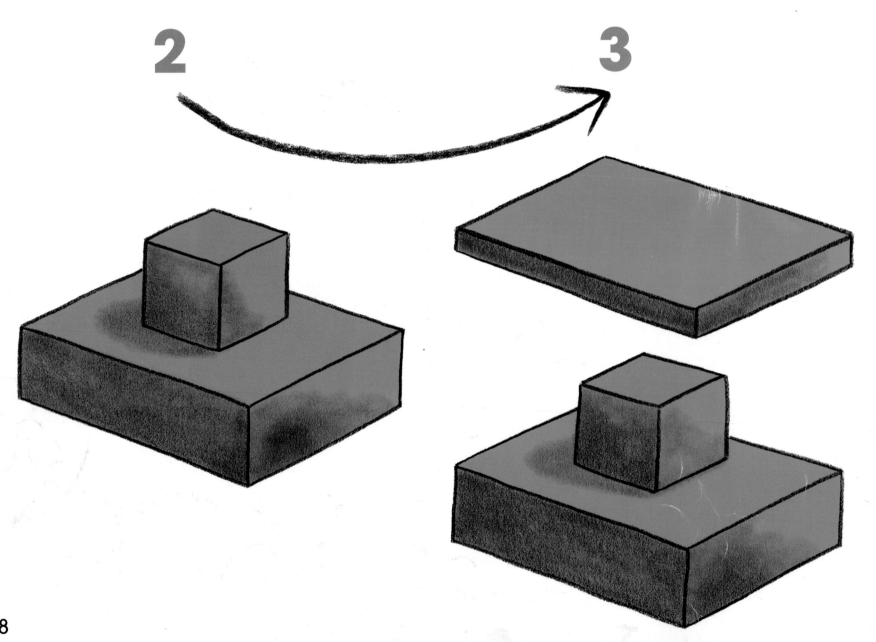

Three blocks can be . . .

. . . a hot dog stand to get a treat.

Jack has

3

Now he'll add two more.

4, 5

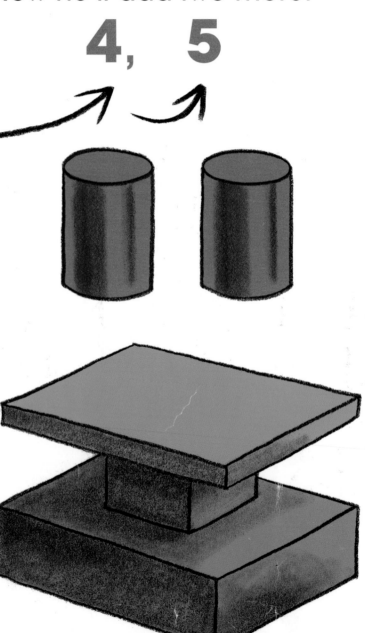

Five blocks will make . . .

. . . a ferryboat out on the sea.

Jack has

5

This time he'll add three.

6, 7, 8

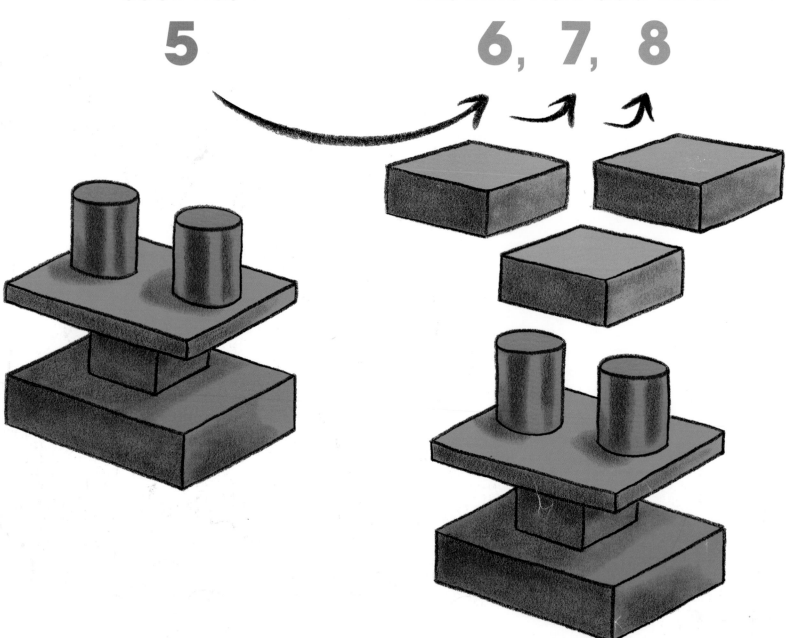

Eight blocks look like . . .

. . . a lookout tower to watch for planes.

Jack has

8

He'll put four more on top.

9, 10, 11, 12

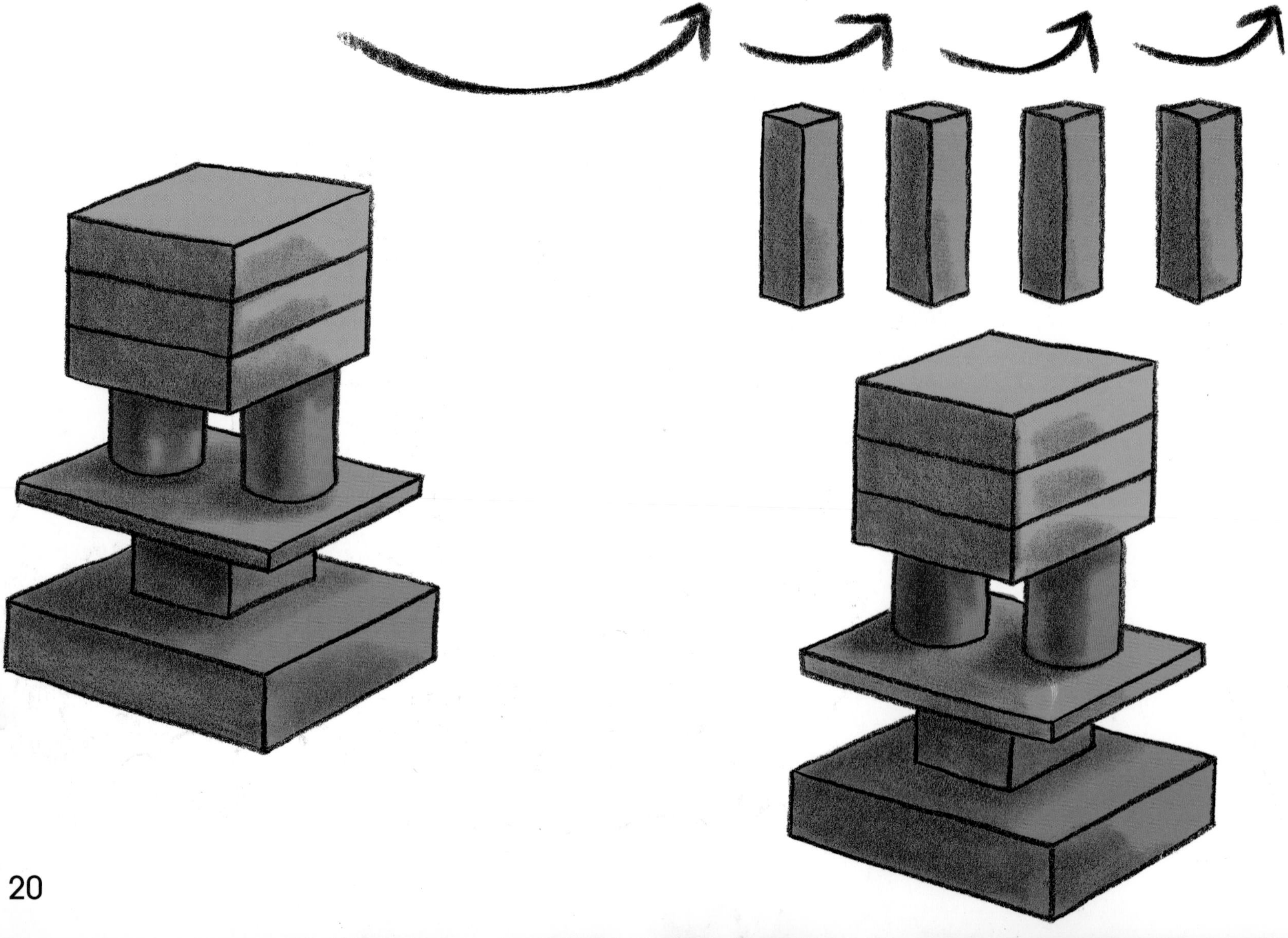

Twelve blocks can be . . .

...the tallest building in the world.

Jack has   Now he'll add five more.

12   **13, 14, 15, 16, 17**

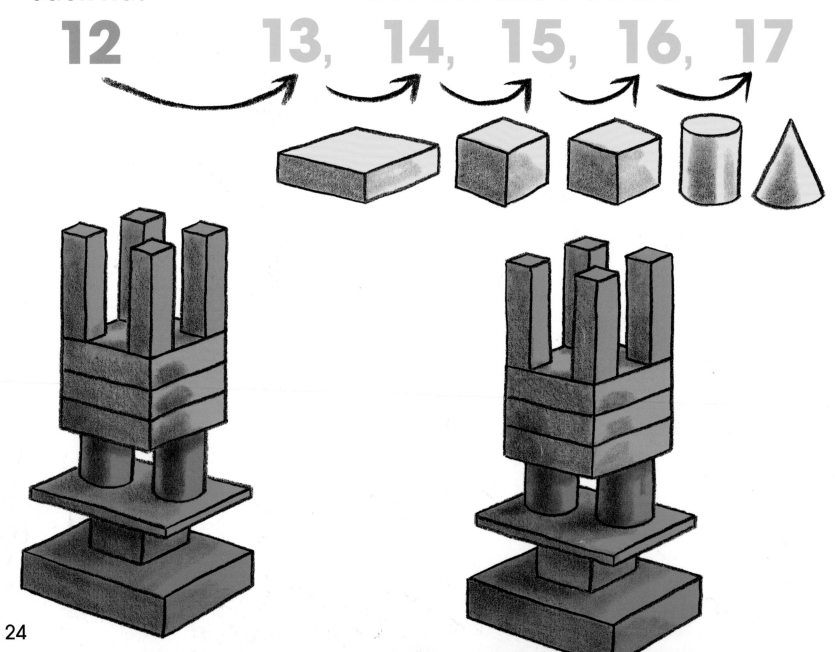

Seventeen blocks become . . .

25

No more rocket ship.

No more building.

28

No more tower, no boat, no hot dog stand.

Not even a robot friend.

Nothing but a pile of blocks . . .

. . . so Jack can start again!

31

In *Jack the Builder*, the math concept presented is counting on, a strategy that children use to solve addition problems. If the problem is 5 + 4, for example, children can solve it by counting by 1s from the 5 to the 9.

If you would like to have more fun with the math concepts presented in *Jack the Builder*, here are a few suggestions:

- Have the child create his or her own *Jack the Builder* experience. Start with 3 blocks and have the child build a shape. Tell the child to add on 2 more blocks and count on from 3 to find out how many blocks have been used. Continue, telling the child each time the number of blocks to add. Ask how many blocks there are now and what each new shape could represent.

- Draw a number line (showing the numbers from 0 to 15) for the child. As you say a number, the child puts his or her finger on that number and then counts on to 15, pointing to each number.

- Continue the number line activity by having the child place a finger on the 5, for example, then count 3 more and give the sum (8).

- Start with 1 block in a box. Roll a die. Use the number that comes up and have the child add that number of blocks to the box, counting on to find the total number of blocks.

- Count on using a calculator. Tell the child to enter a number from 1 to 10 and then enter "+ 1." Before the child enters "=", have him or her tell you what the next number will be.

Following are some activities that will help you extend the concepts presented in *Jack the Builder* into a child's everyday life:

Adding Up Cards: Remove the face cards from a deck of cards. Shuffle the deck and turn over two cards. Have the child count on from the smaller number to the larger number. (An ace counts as one.)

Making Change: Set up a make-believe store. Let the child be the storekeeper. Pick out an item that costs, for example, 27¢ and give the child 30¢. Using pennies, have the child count on from 27 to 30 to make change for you.

Folding Laundry: After folding a few T-shirts, towels, or socks, tell your child how many were folded. Then ask him or her to count on the unfolded items and tell how many T-shirts, towels, etc. were in the laundry all together.

The following books include some of the same concepts that are presented in *Jack the Builder*:

- QUACK AND COUNT by Keith Baker

- SITTING DOWN TO EAT by Bill Harley

- THE VERY KIND RICH LADY AND HER ONE HUNDRED DOGS by Chinlun Lee

33

3 1170 00710 1656